Humanity versus Empire

by Rolf A. F. Witzsche

Text Copyright (c) 2015 - Rolf A. F. Witzsche

Contents

3

4

About the Illustrated Science series
On the Ice Age and Climate Change
and the book

Humanity versus Empire
Book 2 of the series, Big Bang Blow Out

With the Big Bang Creation theory, which defines the universe as entropic and naturally self-collapsing, the universe has been dragged into the political scene as a scientific excuse to justify the existence of systems of empire that collapse all nations and people on Earth who are subjected to such systems that are totally entropic in nature. The theory of entropy, projected onto economics has made everything appear inherently small, finite, and collapse oriented, instead of being future oriented and development oriented. If this deeply intrusive mythology is not reversed, we commit our children and one another to death by starvation, because then the Ice Age Challenge will not be addressed and be responded to, so that the coming phase shift to glaciation conditions in the 2050s will overwhelm humanity. Very few people will survive the consequences. That's why 3 books are needed to explore the issues.

Mainstream cosmology regards the universe, the galaxies, and the solar system exclusively organized by gravitational force that is known to be the weakest universal force. Mass and gravity are all that the Big Bang Theory allows. However, the next higher-order force in the universe is the electric force that is 39 orders of magnitude stronger than the gravitational force. It is expressed in plasma that makes up 99.9999% of the universe. This reality is not allowed to be recognized as an organizing force in the universe, because it is expressed in electrically charged plasma that is deemed not to exist. Cosmology thereby imprisons itself with cosmo-mythologies where nothing is actually true, and humanity becomes imprisoned with it, by the false concepts. One of biggest imprisoning fudge factors, is the Big Bang theory itself.

While technology has furnished astronomy with amazing capacities for looking at the universe, ironically, what is observed is being falsely interpreted on the basis of assumptions that are simply not true, that are mythological assumptions. As a consequence, ironically, mainstream astronomy looks at the universe blindfolded. What comes out of it, of course, are tragic misperceptions. The results are often so confusing that mysterious fudge factors need to be invented to make the results appear plausible. No such fudge factors are needed in plasma cosmology.

With the next Ice Age on the near horizon, potentially beginning in the 2050s, we cannot afford to play games with fudge factors. The recognition of the true nature of the universe, the galactic system, and the solar system, that together drives the Ice Age dynamics, becomes an existentially critical issue. If humanity remains 'asleep' on this front, we may all die in the easy chair of the consequence when the glaciation conditions resume, which evidence promises, will happen quickly.

Plasma in the physical universe is as challenging in perception as the spiritual domain in the human sphere. Both are invisible, except by their effects, but they are understandable and knowable. But how does one break away from the fairy tales that inspire delusions? Answers must be found.

With the Ice Age Challenge now before us, we face two imperatives. One is to understand the real physical dynamics that power and affect the Sun, and with it to create the physical infrastructures that enable human living to continue in an Ice Age climate. The second challenge, and this is the greater challenge, is to raise up our humanity to such height as will impel us to get the job done. Some say that miracles are needed on both fronts. But what of it? Are we, as human beings, not the miracle makers on the Earth?

In the real universe, the cosmic operations are anti-entropic in nature, and expanding and progressing. We, ourselves are evidence of this progression. Should this progression have ended? Neither is our Sun isolated from the progressive nature of the universe, but expresses its dynamics, its resonating plasma streams, and their reflection in the climate on Earth. Shouldn't we develop ourselves spiritually and culturally, likewise?

Climate Change reflects the nature of the universe. It should also be reflected in us.

The Earth itself is the creation of the Sun, with its atoms having been massively synthesized in high-energy times near the center of the galaxy.

The synthesizing plasma fusion is presently at a low state, though it is currently enhanced for our Sun by electromagnetic 'Primer Fields' that focus interstellar plasma onto the Sun in a highly condensed manner. When the plasma-focusing system becomes inactive, below the required threshold conditions, the Sun reverts to a type of cosmic default level with 70% less energy being radiated, and higher rates of solar cosmic-ray flux being experienced.

At the present rate of plasma diminishment being experienced, the solar activity phase-shift threshold to the next Ice Age period may be crossed in 30 years, or in the 2050s, most likely. With the primer-fields system gone inactive by then, the climate on Earth will get 40 times colder than the Little Ice Age in the 1600s had been. Ice core evidence promises that. Without the needed preparations for human living in such an environment, 99% of humanity would die of starvation, both by the cold, and by CO_2 depletion that diminishes agriculture, as more CO_2 becomes dissolved into the sea.

With the 'Primer Fields' being critical for our very existence, the exploration of them is likewise critical.

In the Little Ice Age, between 10% and up to 30% of the populations in Europe had perished by starvation. The last Big Ice Age was evidently vastly harsher. Only 1-10 million people emerged from it alive. That's all we had after 2 million years of development. We want to do far better this time around; and we can, with large-scale technological infrastructures for our food supply. But will we create them? Will we get the job done in the 30 years that we still have left before the Ice Age starts anew? Will we even consider it? And how certain are we that the phase shift to the next glaciation period will begin, as the evidence suggests, in the 2050s? We have no slack on this front. Should we fail us on this absolute front, we would be committing suicide.

Numerous fields of evidence tell us that the next Ice Age is near. That's where the truth begins. Most of the evidence was discovered in the 1990s and thereafter. Some evidence is measured in ice cores; some is measured in space, by satellites. Some measurements are also made on the ground in terms of measurements of the Earth's magnetic-pole drift observed in northern Canada. All of this is seen combined with high-energy physics experiments at a leading national laboratory, and is also explored in the small in static experiments.

So, what will the answer be? Will we move with the evidence? Or will we lay ourselves down to die by default?

It takes an independent researcher to brake the taboos that have kept mainstream cosmology imprisoned, increasingly, during the past century, even while what is regarded as taboo is known to be wrong.

The Illustrated Science series is intended to open the scene beyond the threshold of accepted taboos, to where the actual physical evidence speaks for itself.

The scope of the existential challenge that the Ice Age brings with it, takes astrophysics out of the academic domain and places it into the foreground as one of the most-critical issues of our time. The big Climate Change events that have already worldwide effects are mere fringe effects in the flow of the ever-changing cosmic dynamics. The big effect, when the Ice Age begins anew, promises to be caused by a dimmer and colder Sun. The loss of 70% of the Sun's radiated energy defines our climate future that begins in the near term.

Sure, we can live with all that by creating new platforms for agriculture that are able to operate under Ice Age conditions. But will we do it? The task is enormous. Or will we fail ourselves on this front? We have no reason to allow us to fail. We have the materials and energy resources on hand to accomplish everything that is required for us to continue to live in an Ice Age World. But will we do it? The big question that never goes away, therefore, is; will we develop our inner resources as human beings sufficiently to get the job done, and to get it done in time? Or will we do nothing, ignore the challenge, and condemn our children and one-another to an agonizing death by starvation? That's the choice.

Towards meeting the inner challenge, I have created the epic series of novels, The Lodging for the Rose. And further, towards meeting the science challenge, I have produced numerous research books and several dozen exploration videos that the Illustrated Science series is modeled after. The work is the result of a quarter century of research, for which numerous elements of evidence in related fields came to light during the timeframe of my research.

It is my hope that the work that went into all of these projects will help in some degree - for humanity that we are all a part of - to write itself a ticket to have a future.

High-resolution color images, of the images in this book, can be obtained at www.iceagetheatre.ca

*The Big Bang trap is paraded as brilliant

Entropy / Empire / Doom

Anti-entropy / Principle / Development

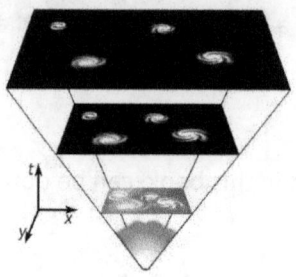

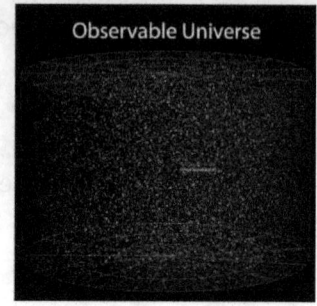

Observable Universe

After the Big Bang start-up
the universe is winding down

A universe that is forever 'beginning'
self-powering, self-expanding without end

The Big Bang trap is paraded as brilliant, even while it is an illusion
that has no substance at its center. However, the Big Bang theory
accurately defines the nature of empire. With this theory, the
system of empire projects its own platform onto the universe, as if
to give its empty platform the appearance of legitimacy, and its
inevitable collapse the face of a natural process, which it says is
fundamental to the universe itself.

With the Big Bang theory

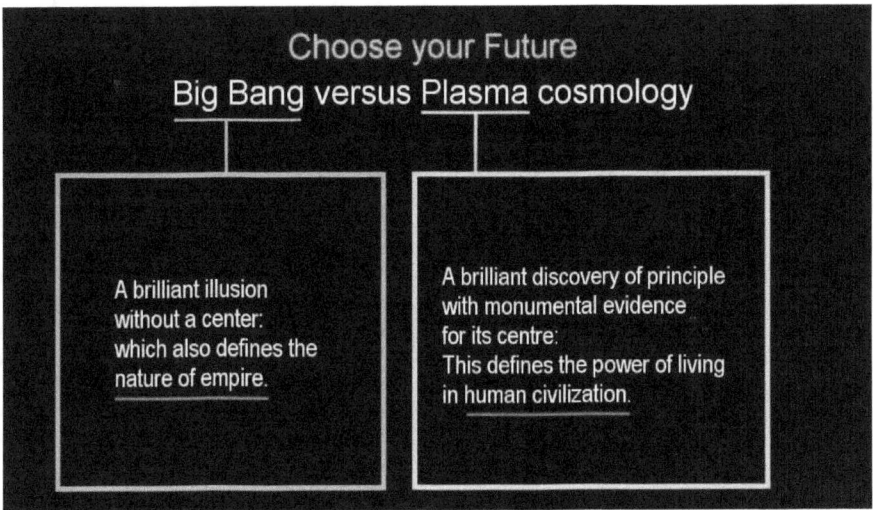

With the Big Bang theory empire says in essence to a struggling humanity, 'the collapse of your civilization is natural. What winds up, winds down again. There is nothing you can do about it. You are impotent, so don't even bother to try. Just look at us: Every empire that ever was, has collapsed. Nobody has yet succeeded in preventing the collapse of empire at any point in history, and of course the collapse of civilization with it. That's built into the nature of the universe. Nobody can alter that. So, just accept the inevitable and lay yourself down to die." This appears to be the modern song.

The lyrics for the empire-song are a lie

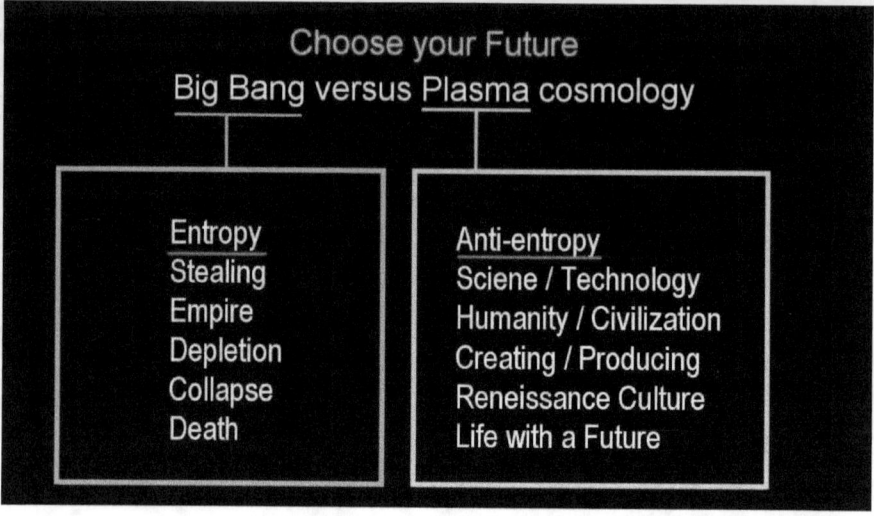

Fortunately, the lyrics for the empire-song are a lie. While it is true that every empire in the entire history of humanity, has collapsed itself without fail, it is also true that every collapse of empire resulted from its inherent emptiness, within, from its devotion to the entropic dream of gaining riches by stealing. In order to prevent its doom, empire seeks to control the world, to keep humanity small so that it will not rebel, but tolerate the pains of entropy, such as providing evermore bailouts to shore-up the thievery.
With these sayings, empire lies to society, big time.

While no empire has ever survived

While no empire has ever survived its defective entropic dream of gaining riches by stealing, whether by trickery, war, or slavery, humanity has been successful a number of times in different places, at evading the resulting doom for its civilization, by human destruction under the system of empire.

Society has freed itself and its future

Society has freed itself and its future, by unlatching itself from empire and its 'corrosive' platform of universal entropy, with a commitment to the general welfare of the nation and nations.

By placing itself onto the platform of anti-entropy

Society has freed its future by placing itself onto the platform of anti-entropy, the platform of the creative renaissance of human development, the platform of cultural, scientific, and technological progress.

On this platform nothing is winding down

On this platform nothing is winding down, except the existence of empire that by its very nature has no place to exist on an anti-entropic platform.

Empire lacks the intention to be creative

Empire lacks the intention to be creative, with its focus on stealing. Empire has never produced an iota of good for the general welfare of humanity. It enforces thereby its own end.

The Big Bang theory needs to be dealt with

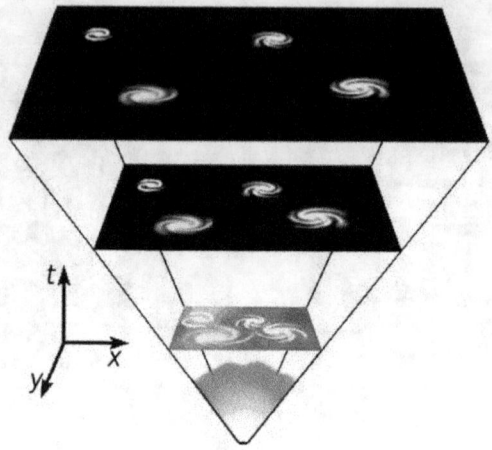

The Big Bang theory needs to be dealt with for humanity to unlatch itself from the entropy of empire, because the theory projects the self-collapsing nature of the system of empire unto the universe. The Big Bang theory is a false theory. It is a fairy tale that has no foundation. It invites society to dream the dream of entropy, the dream of universal depletion.

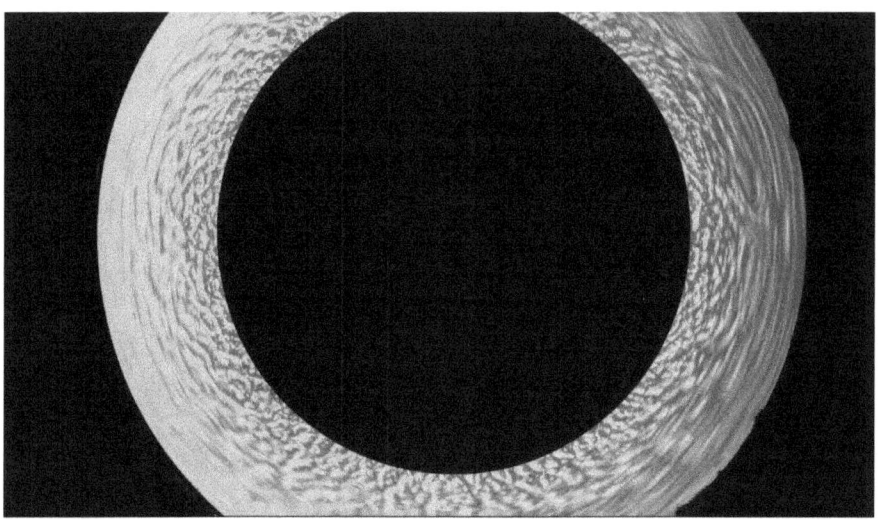

The Big Bang theory, the theory of the inner emptiness of the universe, which it projects unto the universe as if it was real, has the effect that it blocks its opposite, the Plasma Universe, which is real, from becoming recognized. That's where its danger lies. It seeds emptiness, where there is in reality boundless energy in every respect.

The Big Bang philosophy

Entropy / Empire / Doom   Anti-entropy / Principle / Development

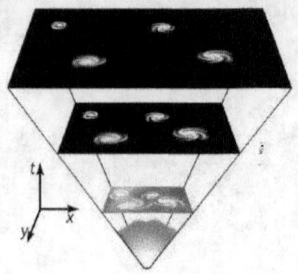 

After the Big Bang start-up
the universe is winding down

A universe that is forever 'beginning'
self-powering, self-expanding without end

The Big Bang philosophy blocks the brilliant discovery of universal principle for which monumental evidence does exist. By society's understanding that the Big Bang theory is fundamentally false, and that its philosophy of entropy is but a tragically empty, false dream, society unlatches itself from the deadening consequences of the dreaming.

Humanity becomes free

By unlatching itself from false dreaming, humanity becomes free to accept the brilliant discovery of the principle of anti-entropy.

The principle of cosmic anti-entropy

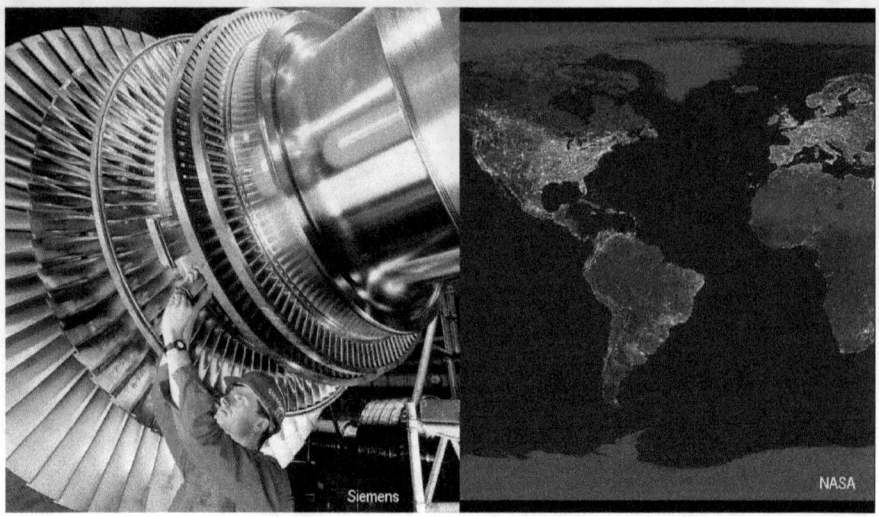

The principle of cosmic anti-entropy is also humanity's own inherent anti-entropy, for which monumental evidence exists that defines the power in intelligent living in a human civilization, which becomes applied inherently scientifically, and progressively with technology, and of course expansively in anti-entropic economics, such as with high-temperature, automated, industrial processes utilizing advanced materials and processes.

The Big Bang theory blocks humanity

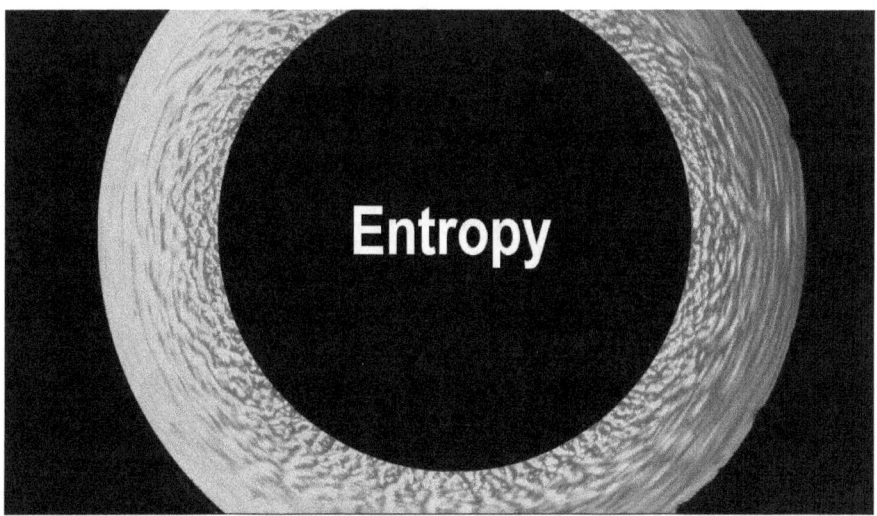

The Big Bang theory blocks humanity from being human. It is a global suicide contract that has been cleverly dreamed up for the masters of empire, as a tool that closes the door on humanity's progressing beyond the empty entropic base that empire dwells on, the entropic base of fuel-based energy production, privately owned, and privately exploited. The Big Bang theory may have been developed in part for the purpose of keeping humanity empty inside and without hope.

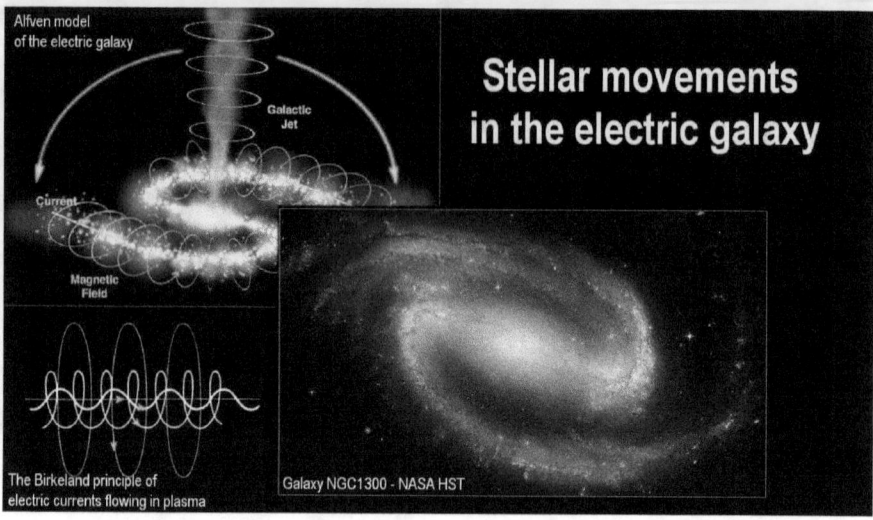

The opposite concept, the universally powered concept, the plasma universe concept, was pioneered in the late 1800s by the Swedish electrical engineer Hannes Alfven.

As primitive as the plasma theory was then, the concept was revolutionary. The promise that it held, for an infinite energy-rich future for humanity, would have rendered the private ownership of the entropic energy resources in the world, obsolete, such as the various fuels. One of the pillars of empire would have fallen by the wayside if the anti-entropic energy concept had been further developed to the stage of implementation. The private energy resources are presently deployed in the massive looting of society by the oligarchic system of empire. And so, Alfven's work, perhaps without him being aware of it, had threatened to pull the rug out from under the system of empire.

Big Bang theory to prevent the collapse of empire

It appears that the Big Bang theory was hastily invented and massively promoted, to prevent the collapse of empire by the development of society from within, which would normally have occurred.

Empire projected its death-model onto the universe

The chokehold on society begins with keeping its center empty, its science. It is said that a great debate had erupted in the 1920s in the halls of empire, between H. G. Wells and Bertrand Russell, over how the danger to empire, of the free development of science, can be prevented. It appears that the promotion of the idea of entropy, projected as the nature of the universe, was the chosen path. With it, the system of empire projected its own inherent death-model onto the universe as if no other model does exist or is possible.

Empire is an empty hole that drains the world

Empire is an empty hole that drains the resources of the world into its empty center, whereby the world becomes darker, and darker, until civilization disintegrates. At this point empire dies with it. Every aspect of empire is entropic in nature by the platform of the universal thievery on which it operates. Not a single empire in the history of the world has not collapsed itself with its own entropy.

The current world-empire is no exception

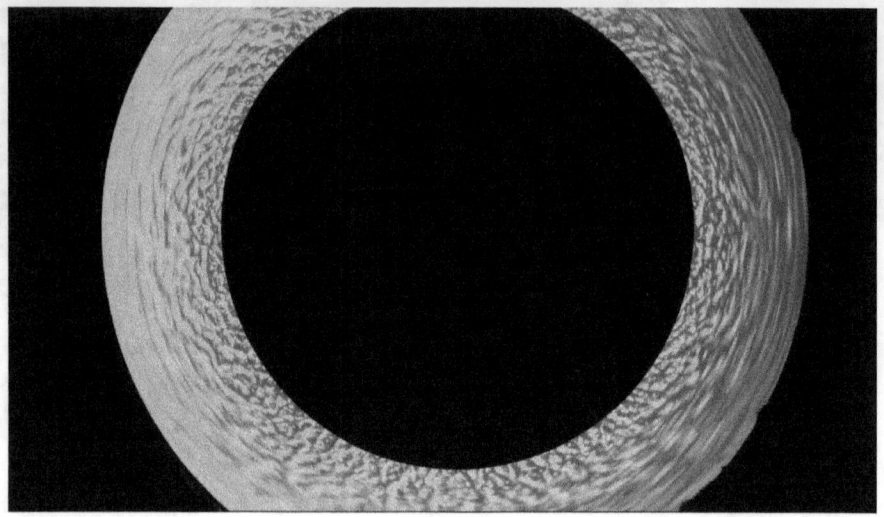

The current world-empire is no exception to the inherent self-collapse of an entropic system. It remains standing in the world as but a ring of smoke with nothing of substance at its center.

Empire is worse than just being empty

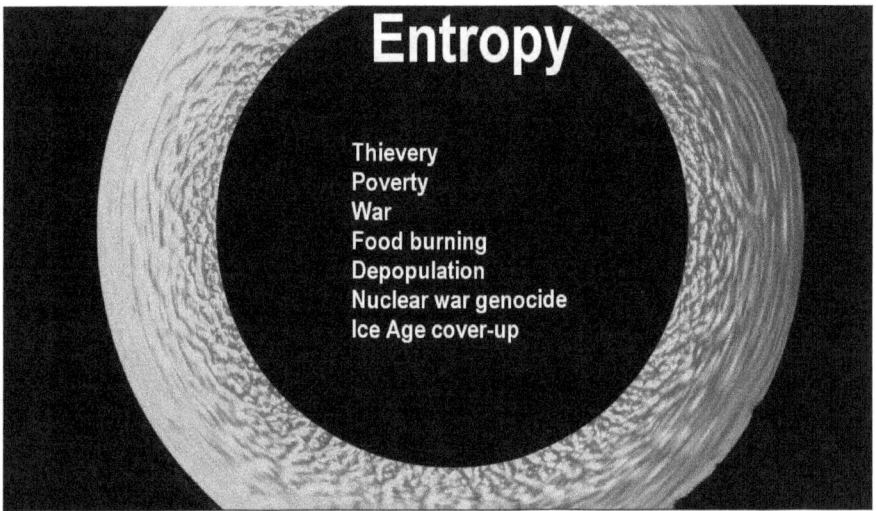

Actually, today's system of empire is worse than just being empty. It is a complex of thievery, poverty, war, food burning, depopulation, nuclear war and terrorist threats. And worst of all its fascist nature blocks the recognition of the near Ice Age that is now before us.

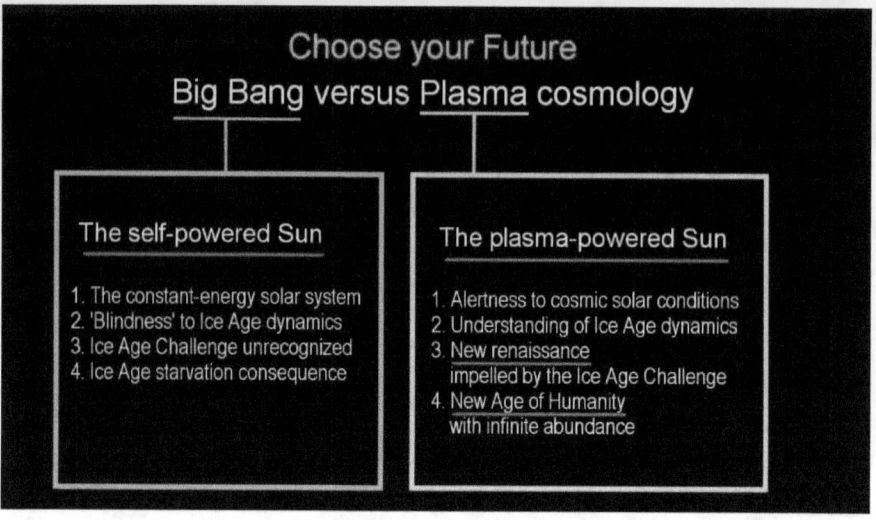

The ultimate form of entropy, built into the Big Bang concept, is its choking effect on science that prevents the recognition of the near Ice Age and its consequences. The extreme lack of scientific recognition prevents the building of the needed infrastructures in the tropics for the future existence of humanity. Under the Big Bang choke-hold on science, 99% of humanity stands poised to starve to death when the Ice Age begins anew, potentially in the 2050s. Only the honest recognition of the science of the plasma cosmology can prevent the consequence that few will survive.

The entropic Big Bang explosion, that theorizes a 'beginning' before which there was nothing, thereby inspires an 'end' for everything. It is a cleverly contrived small-minded concept that appears to have been intentionally staged to induce small-minded, empty thinking, into cosmology as a means for keeping a lid on the progressive development of the plasma-universe concept that the system of empire cannot survive, which as of old will take humanity to the grave with it if humanity remains latched onto this empty system. That's the choice before us.

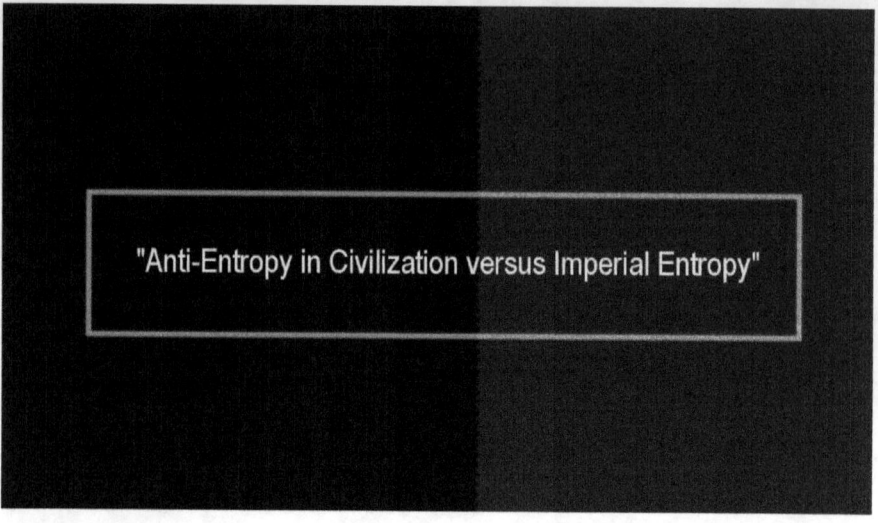

"Anti-Entropy in Civilization versus Imperial Entropy"

The human bang, the real big bang

Should we not rather choose the anti-entropic 'heaven' of grand discoveries that spark a revolution built onto the creative and productive power in society, expanding into all directions? That's where the human bang, is located, the real big bang. This is the choice before us, which the Ice Age Challenge should inspire. This choice would spark an explosion of ideas and subsequent creations. For example, with basalt as the feedstock for a new industrial revolution, and it being shaped in automated high-temperature large scale industrial processes, we have the power at hand to create the needed 6000 new cities with ease, for a million people each, to meet the Ice Age Challenge. And those would be free cities, with free housing for all. Also, the 6000 new cities could all be built in time to enable the relocation of the northern nations prior to their territories becoming uninhabitable. All of this becomes possible when we choose to step away from the entropy of the dark hole of empire.

We can get to this stage of freedom

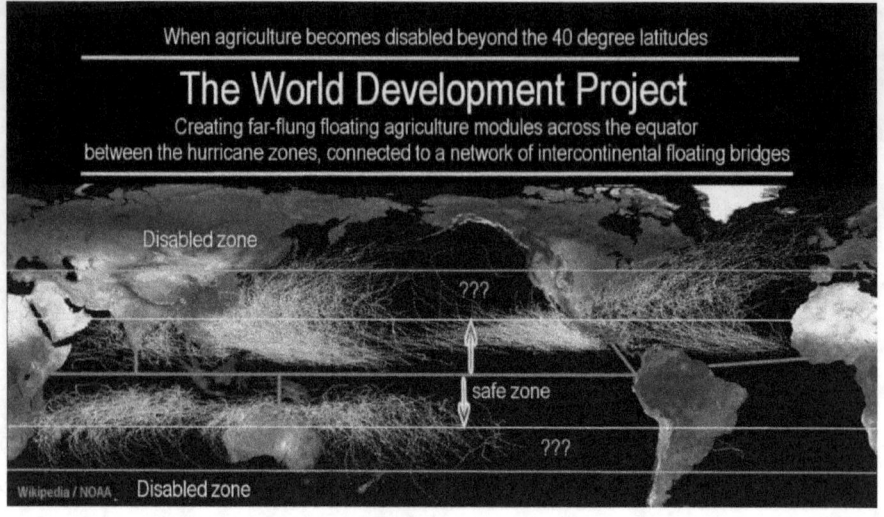

We can get to this stage of freedom relatively easily, by not allowing ourselves to be enslaved to the imperial platform. Once this decision has been made, society will discover its creative power and also relocate all of its endangered agriculture into the tropics before the Ice Age begins, placing it afloat on floating modules made of basalt, interspersed with floating cities that offer universal free housing, and with vast new industries attached that are needed for meeting the human needs.

All of this is physically possible right now, with the already developed technologies and readily available energy resources, for the utilization of the near infinite volumes of the highest-grade materials that are sitting process-ready, unused, on the ground. The vast potential for this type of revolution in human living, which is already critically needed in the modern world, is not being blocked by any physical limits, but is blocked by the 'devil' of entropy in the kingdom of empire that, in its numerous ways, rules the world.

In the shadow of blocking of humanity

In the shadow of this blocking of humanity, civilization is fast breaking down. It is choked by the composite 'devil' of 'empire and entropy' that operates as a single package. The result is the collapse of civilization towards nuclear war, while humanity's vast human potential remains dormant, unrealized, and its future is squandered and is becoming potentially lost.
This tragedy is not what our choice should continue to be, which leads to extinction in several ways.

Our present stage is precarious

Our present stage is precarious, indeed, but it is so only, because the spark in the heart has not yet exploded into the great world-enriching fire that the term, humanity, should inspire. When this spark in the heart will awaken and live up to its inherent promise and light the fire in the heart, our civilization will have a wonderful bright and colorful center that is rich, and beautiful, and substantial, and anti-entropic.

Running away from anti-entropic economics

Society is presently running away from anti-entropic economics, even while it is the only platform that actually exists for building a civilization on. However, society cannot evade the consequences that ensue when it is running away from this only platform.

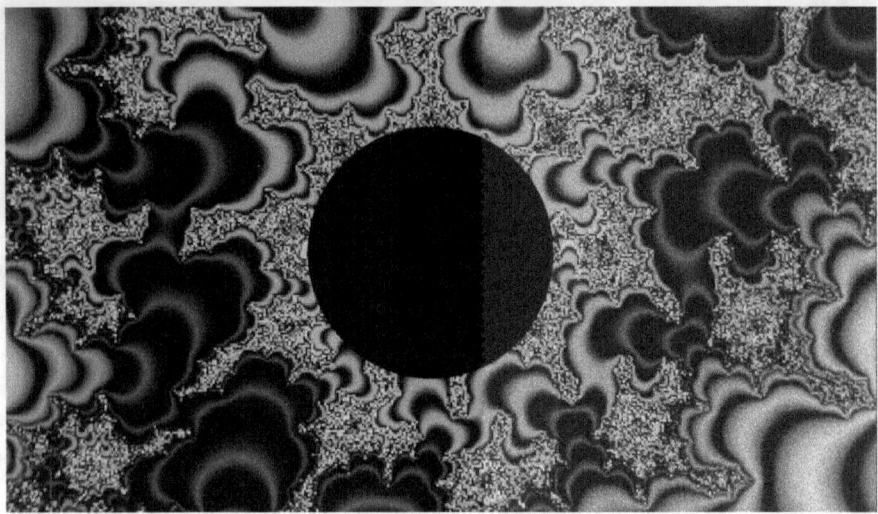

The entropic platform is not a platform for civilization. It is the platform invented by empire in support of its looting operations. Nor is the entropic platform, a platform for economics. It is not designed for building anything on it. It is designed for diminishing everything, for tearing everything to the ground. It is the platform for the money bags, that's required for sealing and for financial derivatives that have the effect of a nuclear bomb in economics. The entropic platform is a power tool for looting that wrecks everything, in contrast with the anti-entropic platform that builds everything.

Two Opposing Platforms in Economics

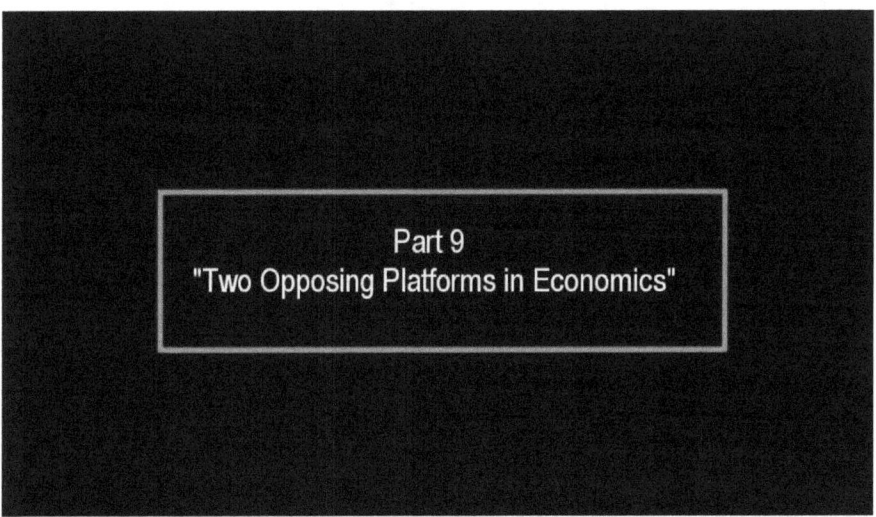

"Two Opposing Platforms in Economics"

The anti-entropic platform

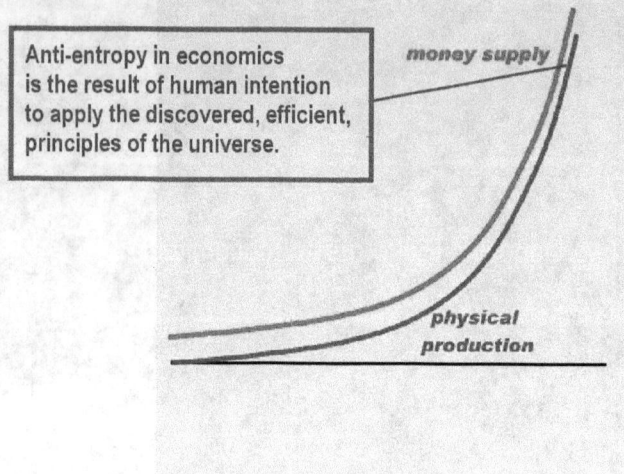

Anti-entropy in economics is the result of human intention to apply the discovered, efficient, principles of the universe.

money supply

physical production

Let me illustrate what an anti-entropic platform looks like, in principle, that is for building a civilization.

The anti-entropic platform is a platform for creating value for human living, by applied human ingenuity and productivity. The anti-entropy, the expansion of value in an economy, is in this case the product of intention. The intention is, to apply the discovered principle of the nature of the universe, to the building of a civilization. The principle that the universe is built on is evidently the most efficient principle. On the resulting platform of intention, society provides to itself whatever financial credits, material, and human resources are required to fulfill its specific needs. Nothing is borrowed here. Nothing is stolen. Every need is met by creative and productive processes. The result gives money its value.

Note, that the factor of intention is of critical importance here. Anti-entropy in economics is the result of human intention to apply the discovered, efficient, principles of the universe, to human living.

A bridge across the tropics

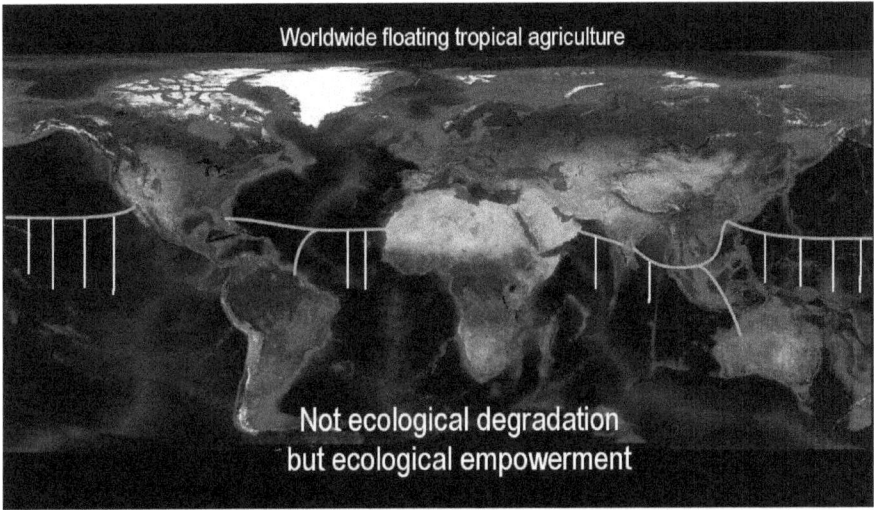

For example, if a bridge across the tropics, with floating agriculture attached to it, is required for society to have a rich and secure future, then a national bank will be founded that provides the financial credits to get the job done. When the job is done, the obligation by society to itself is extinguished. The credit that a national bank would utter for such a project, would not be regarded as a debt to be repaid, because the value created for society, by the completed infrastructure, has at this point already extinguished the obligation by society to itself. That's revolutionary, isn't it?

Credit to get an enriching job done

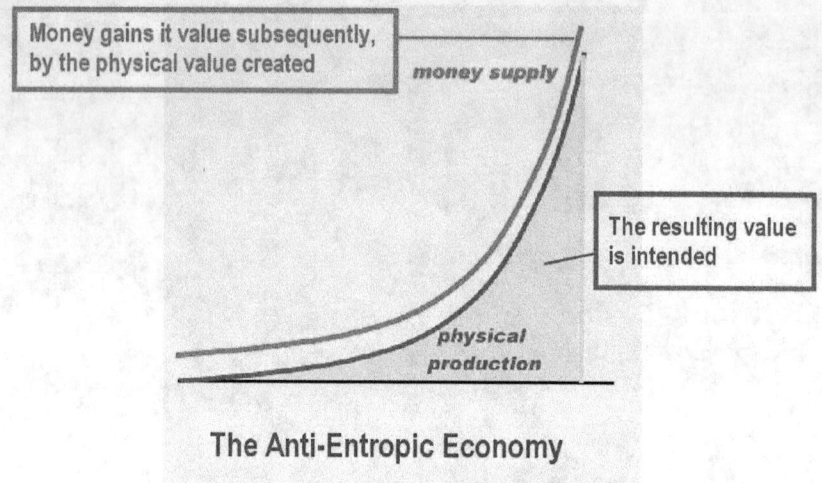

The Anti-Entropic Economy

If money is created as credit to get an enriching job done. The resulting value is the object that is intended. Money gains its value subsequently from that. It gains its value from the physical value created that reflects itself in a richer life in society. The richer life is represented by the shaded area. Which is the intended value.

The process may not even involve the creation of money as credit. Other forms of compensation may be applied to get a job done. The focus is on the intention for physical value to be produced for society, whereby its living becomes richer.

The anti-entropy in this process is not located in money itself at any point, but is located in the physical value created by the human creative and productive process. The entire process is driven by the intention to meet the human need, and secondarily by the intention to provide the needed credit, or promise, or commitment to get the job done.

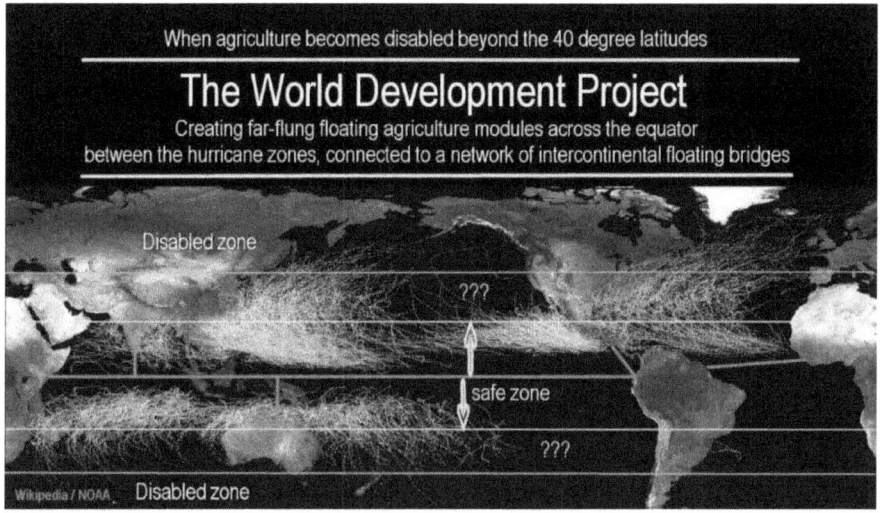

This type of economics in civilization will never be focused on what it will cost to create a richer world. It will be focused on the value that can be created for civilization. The key question will then become, what else can we build for us to make our world grander, our living richer, and our existence more secure?

And so it will be, that increasingly more and more will be built, for ever-greater value being created for society.

On this anti-entropic platform that increases to productive power, the Ice Age Challenge will be met with ease, and may even become a sideline issue. The financial credits that may be uttered to get the job done, will of course never be repaid, as the debt has become extinguished when the job is done, that is when the objective has been fulfilled.

The revolutionary principle - that the debt that society owes to itself to create world-enriching infrastructures, will be extinguished when the infrastructures are completed - totally eliminates the private estate monetarist platform. No money bag will ever lend a penny on the platform that nothing will be paid back in money, nor any interest that it seeks for profit.

*The money-bag system

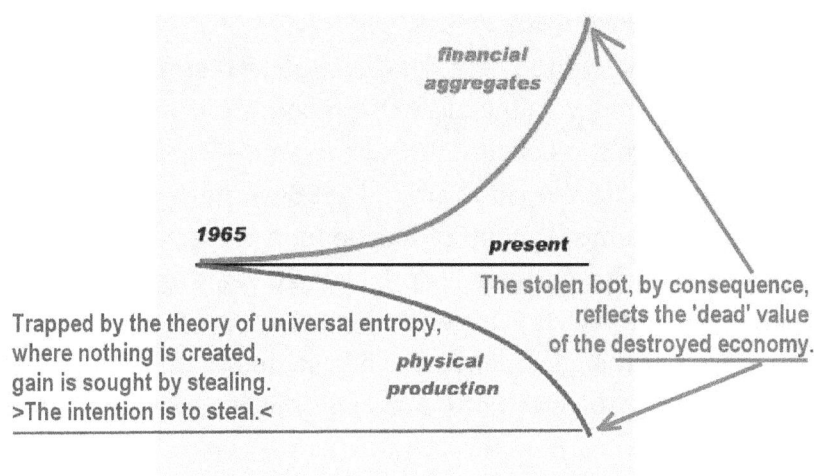

financial aggregates

1965

present

Trapped by the theory of universal entropy, where nothing is created, gain is sought by stealing. >The intention is to steal.<

physical production

The stolen loot, by consequence, reflects the 'dead' value of the destroyed economy.

The money-bag system has a different intention standing behind it, than to seek the development of humanity. Its intention is to swell its bags, instead of producing value for society. The intention is to steal, instead of to produce. The intention renders the money bag system entropic in nature. Trapped by the theory of universal entropy, where nothing is created, gain is sought by stealing, and so on. This characteristic applies to almost all forms of monetarism, from the stock markets wagers, to financial derivatives gambling, to international currencies speculation, etc.. All of these markets are huge, where nothing is produced, though enormous profits are drawn.

When profit is claimed where nothing is produced in value, the profit is essentially stolen from the physical living of society. However, the processes of stealing, which diminishes the productive power of society, as the graph shows, also diminishes the value of the stolen loot, regardless of its volume, as it stands as a claim against the product of the collapsing economy. The stolen loot, by consequence, reflects the 'dead' value of the destroyed economy. The bottom line is, that the widely held belief in the false theory of entropy, has unavoidable entropic consequences. And those are tragically real.

That's what is illustrated here. The illustration presents in principle the entropy inherent in the oligarchic monetarist system. Every empire is tied to this collapse function by its very nature. The illustration was developed at around 1995 by the world-renowned American economist Lyndon LaRouche for a presentation at the Vatican. LaRouche illustrated in principle 20 years ago the dynamics that the western world is experiencing today. Our tragic experience is the result of the general belief in a false theory that has become accepted and assumed to be beneficial, while the opposite is the case.

Empire is doomed by its own premise

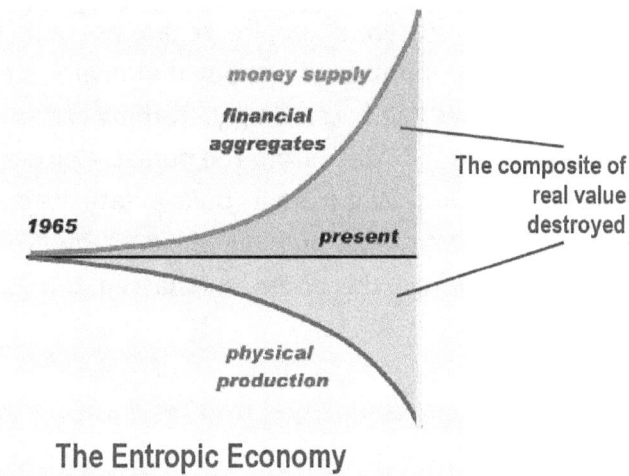

The Entropic Economy

The shaded area between the two curves, represents the destroyed economic value by the collapse process. The destruction of value is typical for an oligarchic, imperial, monetarist system. No other result is possible in an entropic system that the system of empire is, where nothing is created and everything is doomed to diminish to zero.

Every system of empire has doomed itself on this platform, for the simple reason that an entropic system cannot be cured, or be maintained.

The system of empire is doomed by its own premise. It is also doomed for the related fact that no money bag will ever regard the richer living in society as the proper fulfillment of the terms for credit uttered. The money-bag system demands the pound of flesh it claims for profit. In a living world, this claim will never be paid. For this reason, the bag-money system will simply vanish off the face of the earth in the near future, and be replaced with an anti-entropic love-coin system that produces value for society.

The key difference between the two types of system, the anti-entropic system presented earlier, and the entropic system presented here, is a difference in perceived theories that guides the intention. Every system of empire, regardless of its form, is built on the false theory of universal entropy that invites stealing, which thereby collapses itself accordingly. Nothing can save this system that was functionally dead from the outset. Neither war nor depopulation can save it from its built-in fate. Its doom lays in itself by the entropy that it has built on. Society, however, has the power to unlatch itself from this doom, and hasten it.

The consequences are world-destructive

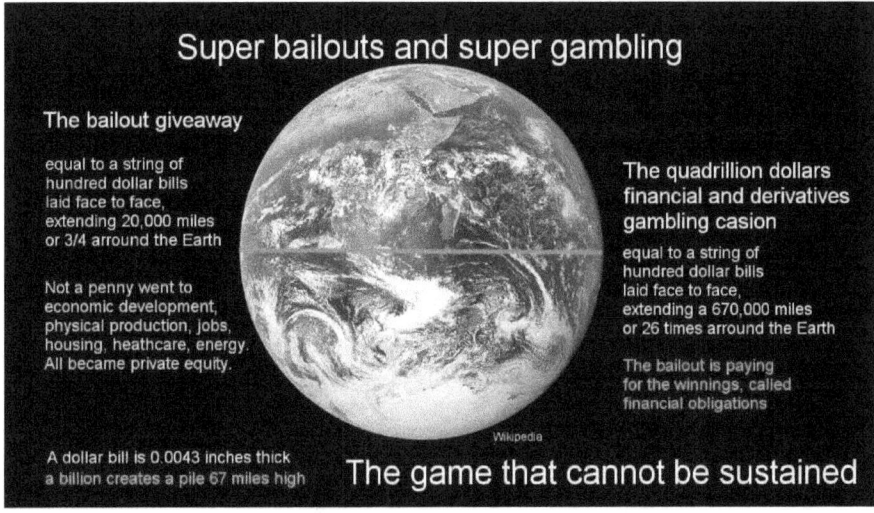

For example, in a gambling arena where 1 to 2 quadrillion dollars are riding the dice, the consequences are world-destructive. For civilization to survive, the entire entropic gambling system needs to be scrapped. And this means, scrapping the system of empire in its entirety, including every law that stands in support of it, without fail.

The anti-entropic system of economics

The anti-entropic system of economics, which reflects the natural creative principles of the universe and man, offers life instead of doom. In its context, science and cultural development are promoted as elements of the anti-entropic system, because science and culture are the key driver for increasing the productive power of society. Real science is anti-entropic. It unfolds the quality of humanity that is without limits.

It was asked in ancient times; how can one know what God is? The questioner was asked to look at the tip of his finger, who does it point to? It points to me. There is your answer, the questioner was told. God is reflected in man. You know no more of God than you know of yourself. So don't belittle God in reaching for the maximum of good, and don't belittle your power in achieving that maximum of good. Here, science begins. With science as a part of you, you can begin to see the universe as it is. And what you see has no limits. The universe is self-powered, expanding, unfolding, and developing itself. And that's what you see when you look at the point of your finger. You see your humanity as the image of God. You see yourself as a part of the process in which you are empowered, and are expanding your proof in the world that you have lived and have developed beyond your wildest dreams.

Free high-quality housing, health care, and education, are factors in this process of unhindered advance in science and culture. Free high-quality housing, health care, and education, are not seen as liabilities in an anti-entropic system, but are seen as opportunities for advancing the creative power of the human being, and thereby of society itself as a whole, which should be promoted and be developed to the highest degree, to maximise human living. In an anti-entropic economy, slum living, poverty, unemployment, miseducation, and other forms of deprivation, stand as economic crimes against society, because they tally up to the deadliest waste possible, the waste of the most precious asset that a society has, which is itself.

The question may be asked here, "Will we get to this point where entropy in economics, and with it, empire, are history?" Of course we will get there, because nothing less will enable humanity to meet the Ice Age Challenge in the few years that we have still remaining till the next Ice Age starts.

We are the supreme being on Earth

Photo by Scott Williams

The Supreme Being

wikipedia

We are the supreme being on Earth, second to none. If we won't inspire us to recognize our infinite potential, nothing will, and we will remain doomed. Thus, the present realization of our infinite future is not blocked by any physical limits of the universe or limits on the Earth, but is blocked by the smallness at heart that society all-to-often allows itself to become trapped by. In this the poets have been correct, who have said in countless different metaphors that the key to the heavens lies in ourselves. It truly lies there, in every respect.

Humanity is anti-entropic in nature

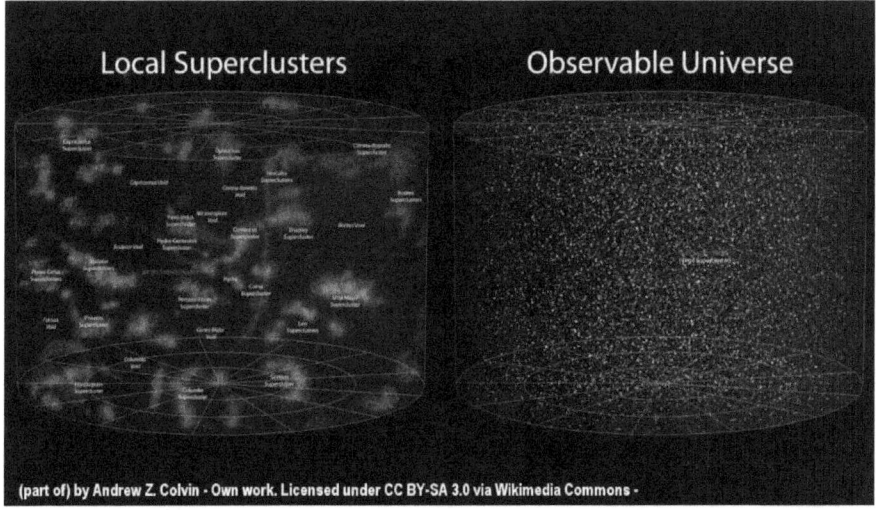

Humanity is anti-entropic in nature and lives in an anti-entropic universe. The numerous forms of real evidence of the anti-entropic nature of the universe, all disprove the Big Bang theory.

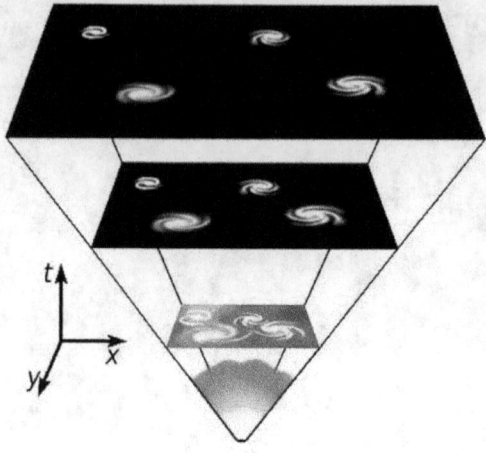

The evidence blows the theory away as a bubble without substance.
Society's illusions about the universe have never changed the universe itself, they have only affected the way the universe was perceived and its principles applied.

*In the Big Bang dream

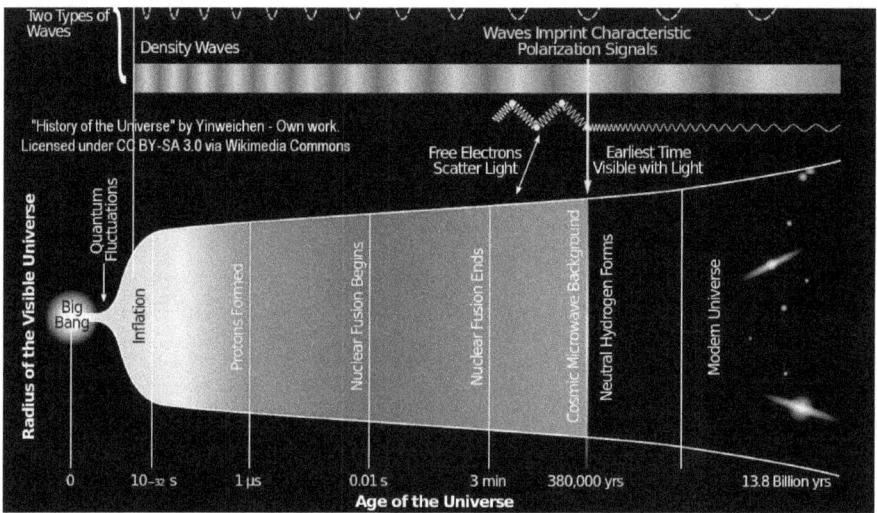

While there is nothing of substance to hold on to, in the Big Bang dream, its forest of confusing illusions is nevertheless, still, rather dense. Consequently, society finds it difficult to trace its way out of the forest to the recognition of the real principles of the universe.

Whether humanity achieves the rich future that it has within reach, inspired by advances in science based on real evidence, or whether it dies of starvation in the forest of illusions when the next Ice Age begins in potentially the 2050s, depends on the choice that society clings to in the present.

The kind of choice that Tolkien places before society

That's the kind of choice that Tolkien places before society in his saga, The Lord of the Rings. He aims to inspire an uncompromised choice.

Unfortunately, our world is darker

Unfortunately, our world is darker, than Tolkien's fictional world.

Our civilization is choked with false theories

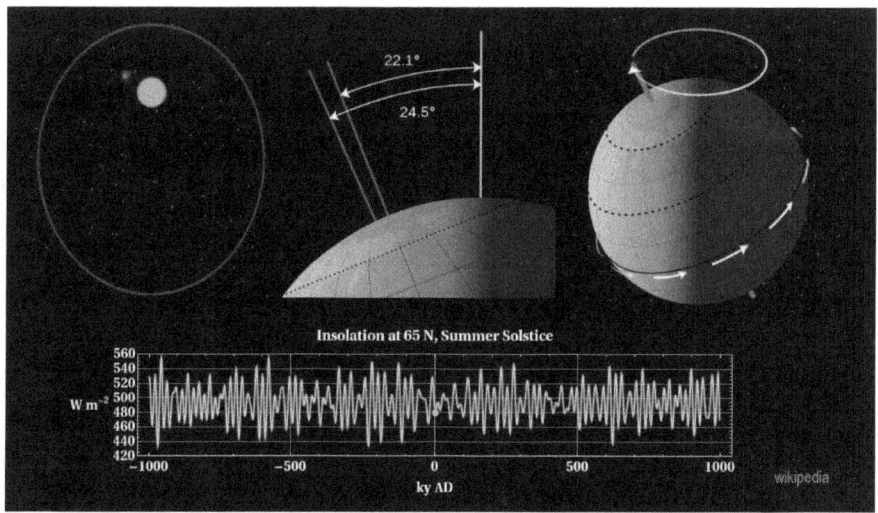

Insolation at 65 N, Summer Solstice

Our civilization is choked with a wide range of false theories that are all pure illusions, and often dangerous illusions for which no supporting evidence actually exists, such as the 'mechanistic Ice Age' theory, where gravity rules everything.

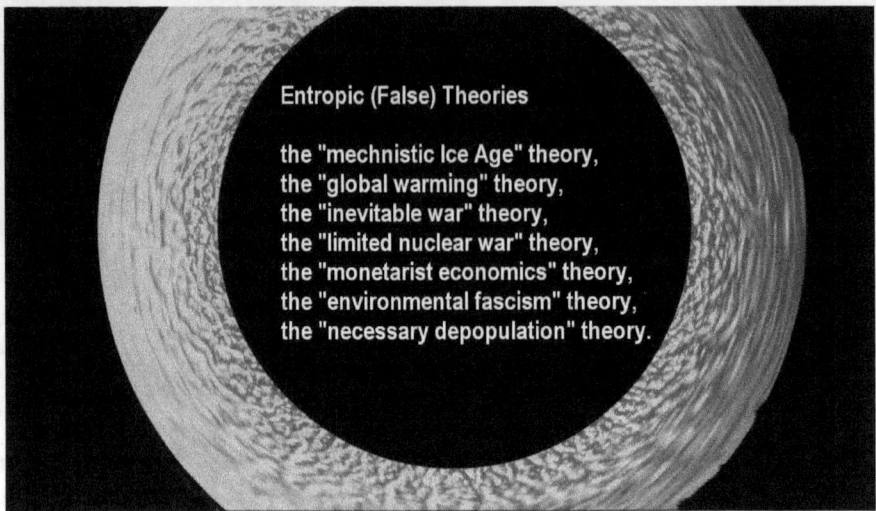

Entropic (False) Theories

the "mechnistic Ice Age" theory,
the "global warming" theory,
the "inevitable war" theory,
the "limited nuclear war" theory,
the "monetarist economics" theory,
the "environmental fascism" theory,
the "necessary depopulation" theory.

Add the 'global warming' theory, the 'inevitable war' theory, the 'limited nuclear war' theory, the 'monetarist economics' theory, the 'environmental fascism' theory, and not least, the 'necessary depopulation' theory, and so on. There is no truth in any of these theories. These insane types of theories are all built up as false concepts at best, and are typically destructive lies by intention.

War is not a natural element of humanity

War, for example, is not inevitable. War is not a natural element of humanity that is rooted in the human heart and soul. The opposite is true. War has never benefitted humanity in any way. It is destructive, and therefore entropic. It collapses civilization. It leaves behind an empty landscape, a giant wound that takes a long time to heal.

Monetarist thievery has never created a nobler and stronger society

Monetarist thievery has the same characteristic. It has never created a nobler and stronger society, but has always collapsed civilization from within.

limited nuclear war' is a false theory

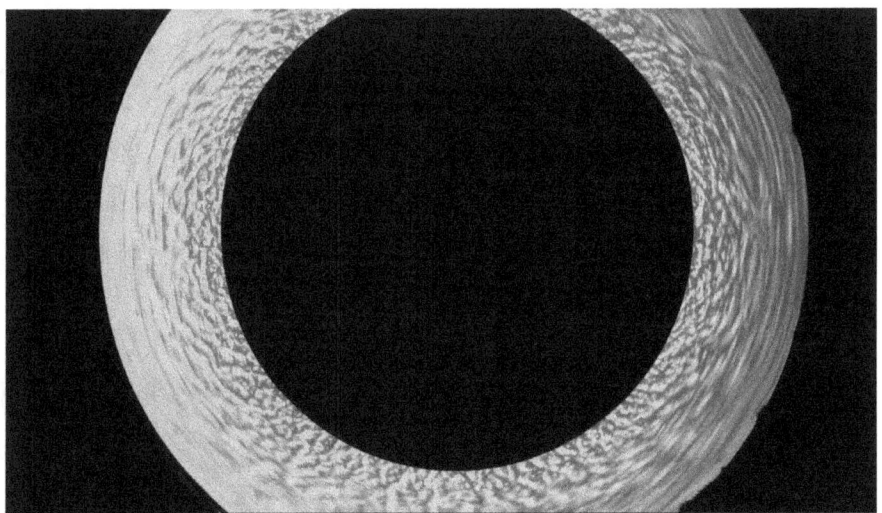

It destroys its own economic center. Neither is the much theorized 'limited nuclear war' possible. It is a false theory. All studies have shown that nuclear war becomes rapidly global, once it starts, with the near-certain extinction of humanity in the wake of it.

Extinction is the ultimate of entropy. But how do we prevent the ultimate?

From entropy to anti-entropy

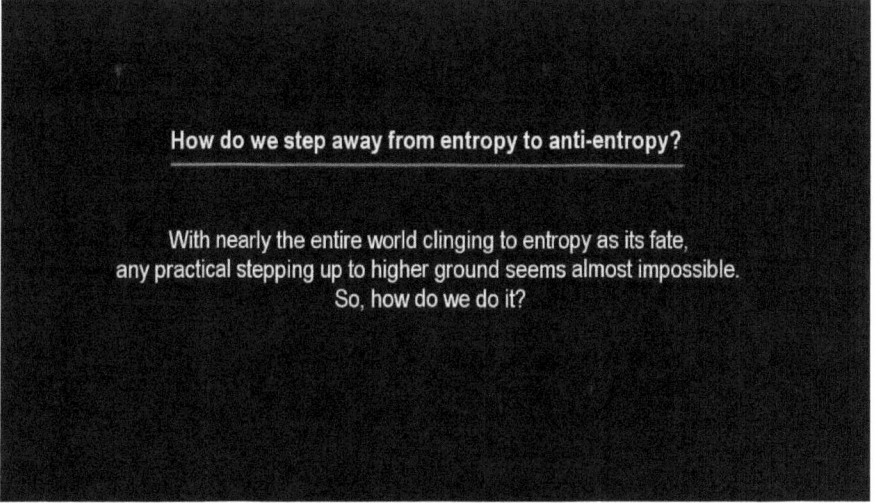

How do we step away from entropy to anti-entropy?

With nearly the entire world clinging to entropy as its fate,
any practical stepping up to higher ground seems almost impossible.
So, how do we do it?

How do we step away from entropy to anti-entropy?
With nearly the entire world clinging to entropy as its fate, any
practical stepping up to higher ground seems almost impossible.
So, how do we do it?

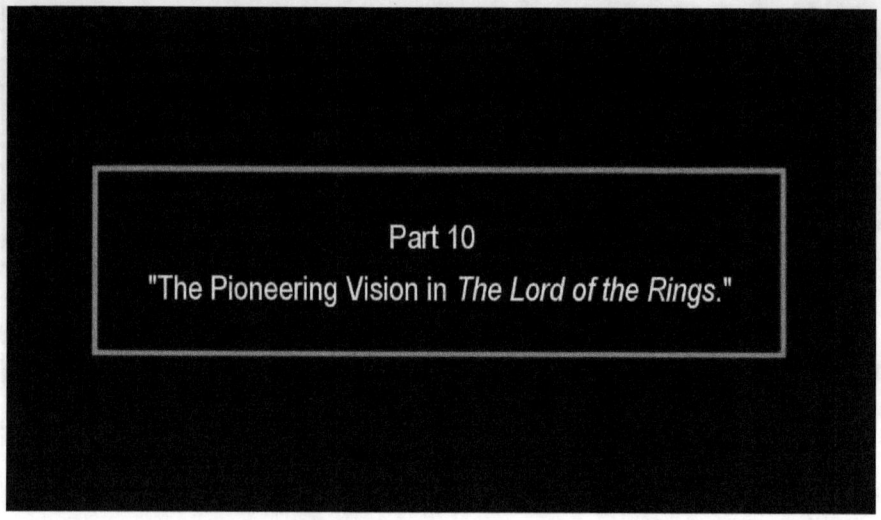

"The Pioneering Vision in The Lord of the Rings."

*Tolkien takes us on a fictional journey

"Gentlemen are rare among the superiors, and even human beings rare indeed."

"The most improper job of any man... is bossing other men. Not one in a million is fit for it, and least of all those who seek the opportunity."

"By 1918 all but one of my close friends were dead." - Tolkien

J.R.R. Tolkien
aged 24 in 1916

Lancashire Fusiliers in a communication trench near Beaumont Hamel, 1916. Photo by Ernest Brooks

The famous author, J. R. R. Tolkien, explores this very question in his epic saga, "The Lord of the Rings." He explores the path of defending a way of life that already reflects the principle of anti-entropy, but which is threatened to become extinct by the corrupting force of entropy that every system of empire is founded on and serves, which throughout history has stood as the most destructive force in the world, and still does so. Tolkien takes us on a fictional journey in an imagined world to explore the dynamics.

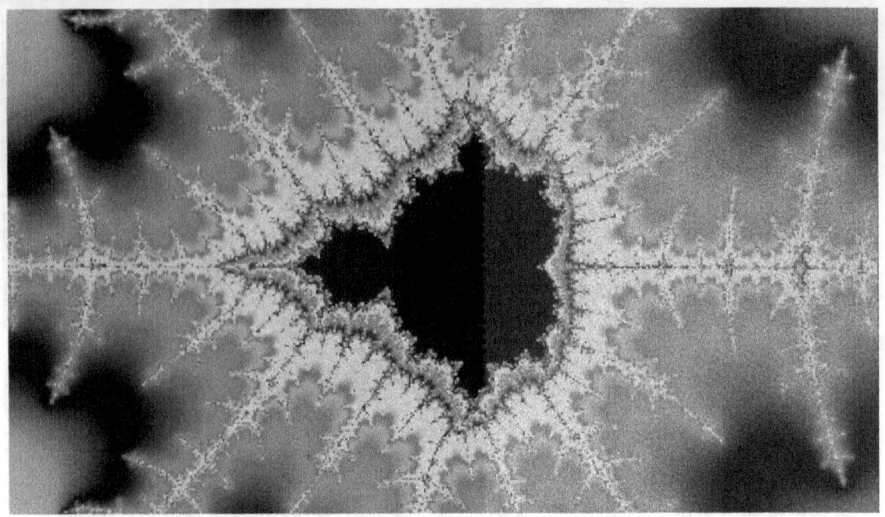

For The Lord of the Rings, Tolkien creates am identifying symbol, the symbol of a ring of fire with an empty center. In the saga, the ring of fire that surrounds nothing signifies the nature of the supreme master of entropy, the master of the high tower of fascism, named, Sauron. The world of men is threatened with extinction by the corrupting force of the fire of the ring of entropy, a type of ring of war that consumes civilization.

Tolkien speaks in metaphor about empire and its effects. In his saga, the ring of fire with an empty center symbolizes the greatest form of evil that has ever established itself, that threatens to end the age of man. The threat defines the battle in the saga, a battle for a society's survival against incredible odds.

Tolkien wasn't far off the mark from the present, in his tale. The modern system of empire seeks the mass-depopulation of the planet from the now 7 billion people living on Earth, to less than 1 billion. The battle in the Ring saga is essentially the same as our battle. But who is fighting the battle in the real world? Who is fighting the forces of entropy?

Lesser rings, rings of gold

by Jeff Belmonte from Cuiabá, Brazil - Flickr. Licensed under CC BY 2.0 via Wikimedia Commons -

Tolkien also speaks about some lesser rings, rings of gold that have corrupted many, with one special ring among them, a "master ring," which the master of the empire of evil had once on his finger, but which became lost for a long time. As a symbol, Tolkien shrank the great ring of fire with an empty center into an object that a person can wear, a golden ring with an empty center that drains a person's humanity away, who thereby becomes a slave to it. The corrupting power of the "master's ring" is so intense in the tale that any wearer of it becomes invisible to everyone around in the normal world as if the wearer did no longer exist. That's what happens in the real world to those who surround themselves with the corrupting influence of entropy. In the saga, the master ring is also called "my precious."

This is the type of effect in the real world that the Big Bang theory has on all who 'wrap' the lie of entropy around their finger. These are they whose contribution to science has essentially vanished as if they had exited the universe, or had never lived. To the people stuck in this trap, the Big Bang Cosmology is everything from A to Z

in the universe, who doom the world thereby. In the saga, for as long as the 'master' ring exists, the world of man is likewise doomed. But Tolkien makes it clear that the 'master' ring, though it must be destroyed, cannot be destroyed with any craft one might employ. A corrupting false theory cannot be destroyed with an axe, but must be returned to the chasms of evil where it was forged. It needs to be traced back to the chasms of empire, the chasm of the fire of lies.

Casting the theory of entropy into oblivion

Tolkien assigned the critical task of casting the theory of entropy into oblivion, to a few pioneers of the society who already live by the principle of anti-entropy in all they do. They live it, by living a rich life with abundant resources that they have created for themselves, and continue to do so. Tolkien placed the tallest task on these few.

Tolkien leaves no room for a compromise

"One of you must win this necessary fight against entropy," a wizard proclaims to a council. Tolkien selects one of the few who already live the principle of anti-entropy, and selects the smallest of them, as if he was saying to him, "only you can do this. You know in your heart what is true. You have lived the principle. The others who are standing on a lesser platform don't have a chance. One cannot win against an evil by fighting on the low corrupting level where the evil has been forged. This doesn't work. But you stand on a higher level. That's why you can win, who are living on that higher level where the supreme principle is experienced as real. This means that You must do this. You have qualified yourself to succeed." Tolkien assigns a wide range of lesser figures in the tale, to help the project to succeed." And they all, cooperating and contributing all that they have, do succeed against the most incredible odds in a fight that spans three volumes in book-form.

Tolkien leaves no room in the story for a compromise with the Empire of Entropy that is poised to destroy the world of man. It is tempting of course, to reach for a compromise, such as to use the supposed 'power of entropy' against its empire who hails it, to defeat it. Such attempts are indeed made in the story, all the way through, and they all fail and result in tragedy.

Tolkien puts the world on notice, not to fail. Still, it did fail in the real world.

The "Glass Steagall" banking legislation

1933 - the Glass-Steagall law to protect that value of the U.S. currency - now repealed

A great attempt had been made to save the USA from the ravishing of the Empire of Entropy. It is known as the "Glass Steagall" banking legislation of 1933. It was designed to protect the nation. However, it was also built on a compromise. Tolkien's tale was written between 1937 and 1949, with the bulk of it during World War II. More than 150 million copies have been printed of the tale. Nevertheless, its fundamental premise has remained unrecognized. The compromise in the Glass Steagall legislation in the USA, has never been addressed, much less overturned, whereby the law itself was eventually defeated, and the protection it had afforded, was lost, with the result of great tragedies for the nation and the world. The tragedies still continue.

Among the foes in The Lord of the Rings

Foreground, left to right: Führer Adolf Hitler; Hermann Göring; Minister of Propaganda Joseph Goebbels; Rudolf Hess

Among the foes in The Lord of the Rings, Tolkien places a group of black-glad kings who were once men, but who became corrupted into becoming creatures without a face and without a soul, who wield destructive power, as they all serve Sauron, the foulest evil of them all whose symbol is the ring of fire with no center.

We have many evil potentates in high places

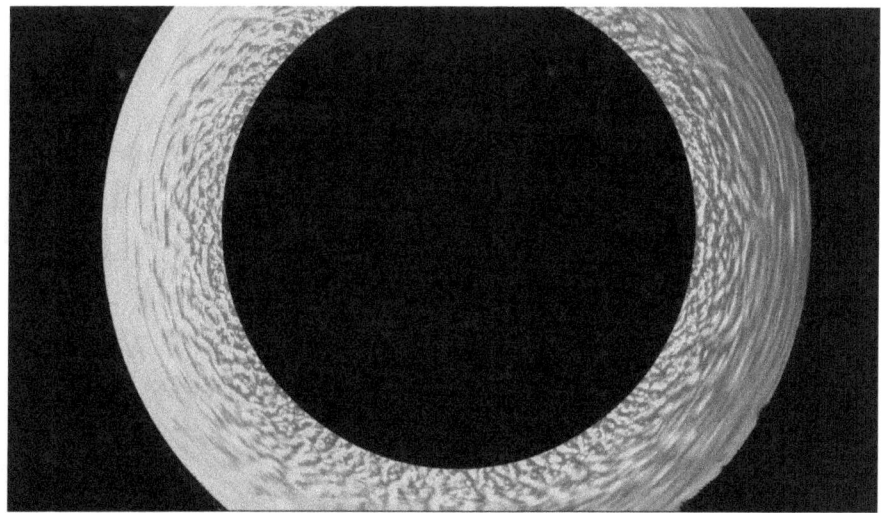

We have many such evil potentates in high places in the world to whom the symbol of an empty center surrounded by a ring of fire, would apply.

Castle Bravo - the first U.S. test of a dry fuel thermonuclear hydrogen bomb - March 1, 1954 at Bikini Atoll, Marshall Islands

Those are the criers for war, even for the big bang of nuclear war, the supernova war that nothing survives.

Masters of false theories

USAF B-2 Spirit
Stealth bomber

20 in service
9,700 Km range without refuelling
Bomb capacity app. 38,000 pounds
Price tag: $2,000,000,000

wikipedia

They are masters of false theories. Nuclear war is a false theory.
Every form of nuclear war is a false theory, even limited nuclear war
that is presently prepared for to eradicate Russia, China, and India.

B-2 Spirit

We see Russia and China placed at the cross-hairs for a surprise limited nuclear war, by scores of empty people who have lost their qualification to be called human beings.

Where people lie to themselves

USA - Ohio-class strategic missile submarine - (14 in service)

In their land, where people lie to themselves, the world-engulfing conflict of limited nuclear war is miraculously deemed survivable in some fashion, for which the stage has been set up, and for what?

When society looses its renaissance of the truth

by Rembrandt
(1606–1669)

When society looses its renaissance of the truth, insanity rules the landscape, whereby the most precious we have on the Earth, which is humanity itself, is doomed to suffer extinction.

"The American Paradox"

America stands as a paradox that has not yet been resolved. It was founded as a republic. A republic is inherently anti-entropic in nature. When forms of government by a society of itself, are focused on increasing the wealth of its living, and thereby its freedom from limitation by the creative and productive power of its humanity, then the resulting system of economics is anti-entropic in nature. It is so, because the wealth of its living, from food to housing to transportation, is increased by the society's scientific and technological progress, and by its commitment to empower this real wealth-building process by all means possible.

In the resulting environment society functions by supporting one-another on the entire front for its common welfare. The dynamic process includes quality education, health care, increased beauty, expanding love, and an elevating culture. On the recognition of this universal-freedom platform many of the former colonies in North America eagerly joined hands to became a nation.

Why has the nation fallen?

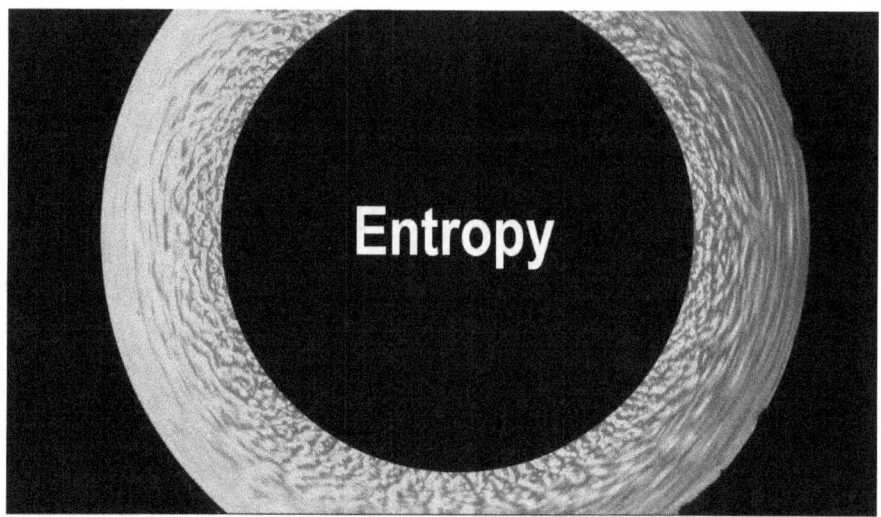

But why has the nation fallen from the high platform it has been built on? Why has it fallen back to the level of the kingdoms of entropy where human value decays?

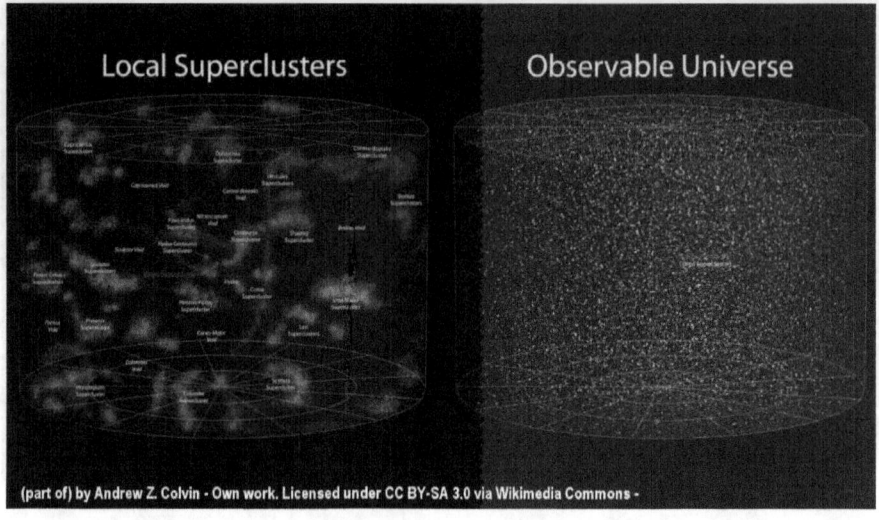

(part of) by Andrew Z. Colvin - Own work. Licensed under CC BY-SA 3.0 via Wikimedia Commons -

America should not have fallen. The universe doesn't fall into entropy. The universe operates on the anti-entropic platform where nothing stands in isolation and is self-consuming, where instead everything is actively powered by a process in which every single star of every galaxy plays a contributing role. Without this contributing, active participation, in the universally creative process that the universe is, with its myriad expressions, not a single star would exist in cosmic space, which would thereby be an empty void.

America the paradox of a fallen star

America stands as the paradox of a fallen star, because a portion of society believes that it can exist and prosper on the opposite platform than that of the universe. The opposite platform, in this case, is the platform of entropy, reflected as the lie in economics that invites the tragedy of stealing from one another, such as by slavery, deceit, greed, and other forms of larceny, even rape, such as by waging war for stealing.

America stands as a paradox, because it aims to exist on two opposite platforms: the platform of Entropy where stealing is King, and the Anti-Entropic platform were society's creative self-development is pursued and is powered by its humanity and its human resources. The conflict between the two systems has shaped American history.

American history is a history of great achievements compromised with great tragedies, a history of America being a light into the world, and of it also being the greatest force of fascism and war that stands poised to extinct humanity on the pathway of universal stealing.

90

The paradox that America became

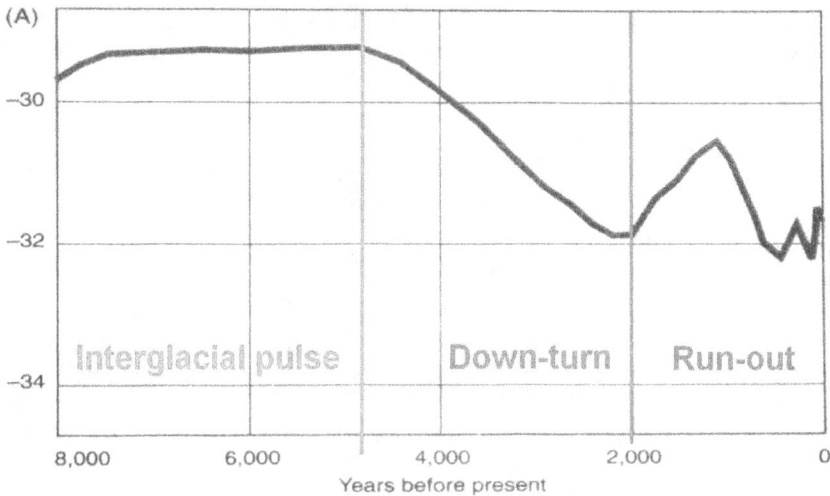

The paradox that America became, appears to have started millennia before its time when the climate of the Earth began to cool from the warm climate of the interglacial optimum. In the shadow of this climatic down-turn, the Earth's primitive biological system became less productive. Food became less abundant. The idea might have emerged that wealth in living can be increased by stealing, which leads to wars for stealing. In the less-productive landscape roving bands may have ravish the lands. Some of the bands became empires - empires that are trapped by the notion of entropy that invites stealing. This is the trap that society fell into, by its small-minded thinking. This deadly trap in thinking still chokes much of humanity.

One of the great masters of the trap

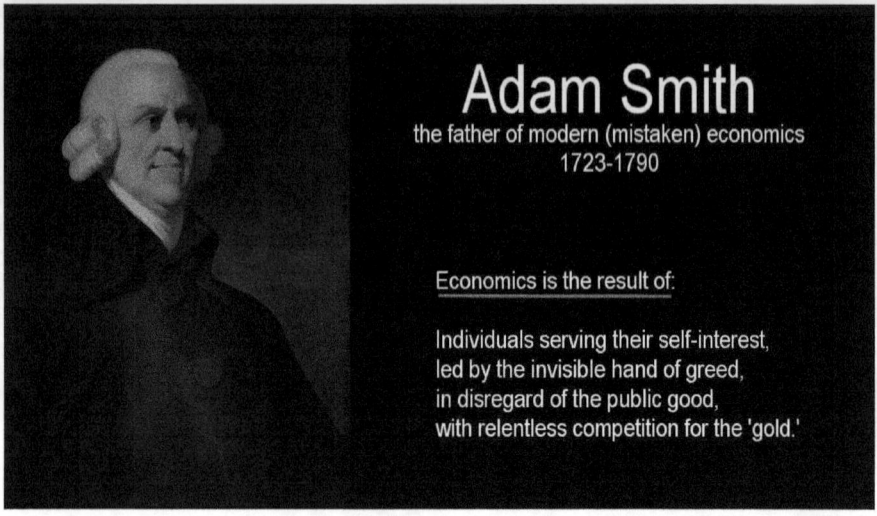

Adam Smith
the father of modern (mistaken) economics
1723-1790

Economics is the result of:

Individuals serving their self-interest,
led by the invisible hand of greed,
in disregard of the public good,
with relentless competition for the 'gold.'

One of the great masters of the trap was Adam Smith, who is hailed the father of modern economics that should be termed, mistaken economics. His take was that economics is the result of individuals serving their self-interest, led by the invisible hand of greed, acting in disregard of the public good.
Smith trapped himself in this prison of entropy where the fierce heartbeat of relentless competition diminishes all human value to zero and elevates money value to be the all-precious. Thus greed becomes a deity that demands the sacrifice of everything else. Adam Smith calls this economics, even while he 'defies' the principle of economics.

Adam Smith was mistaken

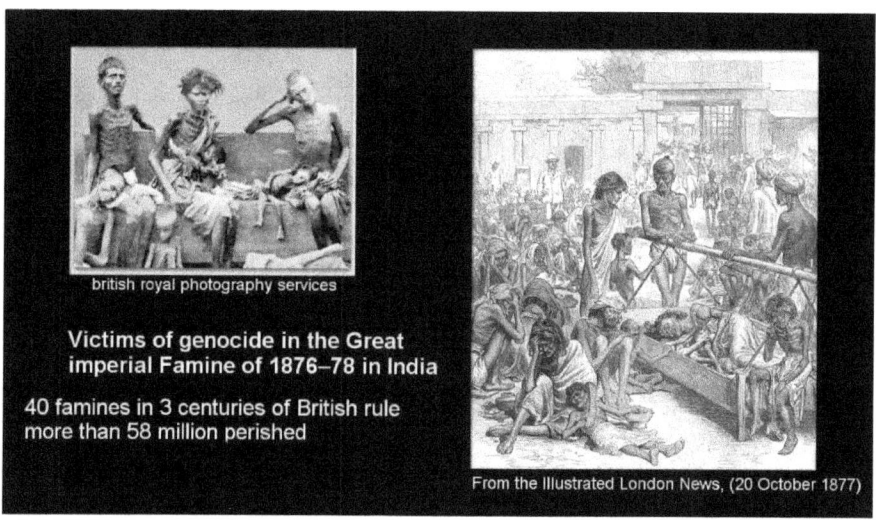

british royal photography services

Victims of genocide in the Great imperial Famine of 1876–78 in India

40 famines in 3 centuries of British rule more than 58 million perished

From the Illustrated London News, (20 October 1877)

Of course, Adam Smith was mistaken. The hand of self-serving greed that acts in disregard of the universal good, is not invisible in the real world, it never has been, but has always been ravishing. Adam Smith stands self-condemned thereby, by erecting a throne that stands opposed to the anti-entropic platform of human development where creativity, science, and productivity are rooted. His throne opposes the only source of true economic value that exists. He opposes the system of humanity where stealing is NOT required, but is regarded as a crime.

The process of stealing as an economic concept, which the empire of western economics has become, is inherent only in the imagined entropic model that emerged out of the ages of small-minded thinking. That's what Adam Smith had idealized, a model of relentless competition for the gold in which human value falls by the wayside.

The Smith model is immensely destructive

When 'money' is for stealing,
wars are waged for increased stealing.

There has never been a war instigated in all history,
that wasn't for the purpose of stealing. Wars are for stealing.

The Smith model is immensely destructive.
When money is an object for stealing, wars are waged for increased stealing. It is useful to note here that there has never been a war instigated in all history, that hadn't been for the purpose of stealing, fundamentally. Wars are for stealing, exclusively.

In the extreme case, as we have it today

Annihilation is assured

500,000 times
Hiroshima
in one hour

The extreme desperation for stealing
leads to nuclear war,
the ultimate 'competition' that no one survives.

Castle Bravo - the first U.S. test of a dry fuel thermonuclear hydrogen bomb - March 1, 1954 at Bikini Atoll, Marshall Islands

In the extreme case, as we have it today in the disintegrating world of imperial monetarism, the resulting extreme desperation for continued stealing opens the gates to the unthinkable, to nuclear war, the ultimate 'competition' that no one survives.

nuclear war,
in any form,
is unsurvivable

Nuclear war, in any form, is unsurvivable.

A dead peace without a human voice

...a dead peace without a human voice

India North (Wikipedia)
Jochen Westermann from München, Germany

Nuclear war will lead to a dead peace without a human voice, or any voice at all.

Far distant from Adam Smith's economics

Mumbai Bridge

"Worli skyline with BSWL" by Woodysworldtv -
http://www.flickr.com/photos/woodysworldtv/5530750545/sizes/o/.
Licensed under CC BY 2.0 via Wikimedia Commons -

The potential future of humanity lies far distant from Adam Smith's economics that still drives the western entropic platform for stealing and destroying, which the developing world is fast moving away from in the emerging cooperative commitment by Russia, India, and China, towards a world of building for the universal welfare of society.

The future of humanity

India Mumbai Bridge (Wikipedia)　　　　　　　　　　　　　Amit Kulkarni

The future of humanity lies in rapidly increasing the wealth in human living. It lies in moving away from destroying humanity, such as by starvation, war, and depopulation, to developing the creative freedom and unlimited productivity that humanity is capable of.

Real economics aims for the goodness in living

"Taj Mahal by Amal Mongia" by amialdia from san francisco
CC BY-SA 2.0 via Wikimedia Commons

Real economics aims for the goodness in living, without stealing, without greed, without war, without destroying one-another in the grasping for gold, but with evermore building, uplifting culture, and beauty. The future of humanity lies with the principle of anti-entropy, the principle of the universe, reflected in the unfolding creative power of humanity.

After India became a free nation in 1947

The Lotus Temple, located in New Delhi, India, a Bahá'í House of Worship completed in 1986. 70 million visitors by 2001

After India became a free nation in 1947, that ended its long history of being looted in numerous colonial processes by numerous masters, in which many tens of millions of people have perished. By claiming its freedom from the thievery, India gave itself a chance to live again, which hadn't been possible during the dark centuries of it being ravished by foreign masters, who had subjugated most of the world at the time, and still do, as they did India.